THE
NEW ICE AGE

THE NEW ICE AGE

FRANKLIN WATTS
NEW YORK/LONDON/1978
AN !MPACT BOOK

BY HENRY GILFOND

Library of Congress Cataloging in Publication Data

Gilfond, Henry.
    The new ice age.

    (An Impact book)
    Bibliography: p.
    Includes index.
    SUMMARY: Examines evidence indicating a
new ice age may be starting and discusses the
social and economic impact of drastic climatic
changes.
    1.   Climatic changes—Juvenile literature.
[1.   Climatic changes]   I.   Title.
QC981.8.C5G54   551.6′9   78–3676
ISBN 0–531–01458–4

Photographs courtesy of

United Press International: pp. 4, 5, and 30; National
Oceanic & Atmospheric Administration: p. 13; Hale
Observatories: p. 22; Yerkes Observatory/University
of Chicago: pp. 31 and 33 (bottom); American Mu-
seum of Natural History: p. 33 (top); United States
Navy: pp. 40 and 43; Maury Solomon: p. 58.

Diagrams by Vantage Art, Inc.

To Maria Rosa and Michael

CONTENTS

ANOTHER ICE AGE?

Are we on the verge of experiencing another ice age? Will the land on which we live, as well as our rivers, lakes, and oceans again be completely covered by a solid sheet of ice, a sheet of ice as much as two miles (3.15 km.) thick?

In the Western Hemisphere the autumn of 1976 was the coldest in ninety years. And almost all of the continental United States suffered record-breaking cold temperatures in the winter of 1976–77. In St. Louis and Cincinnati, cities that usually escape deep freezes during their winter seasons, the temperatures dropped to –13°F (–24°C) and –24°F (–31°C), respectively. Buffalo was buried in record-breaking snowfalls and required the help of cities as far away as New York to free its roads, open its factories and schools, and bring food to its marooned population.

Ice dams, ten feet (3 m.) tall and taller, blocked traffic on the waters of the Mississippi River above the city of Cairo in Illinois. Lake Erie, for the first time anyone could remember, was frozen solid.

Meanwhile, California continued to suffer from drought, and the Midwest was hit by severe wind and dust storms. Both the drought and the dust storms, along with other observable climatic phenomena, may well be indications of changes in weather patterns, not only for the Western Hemisphere but for the entire world.

And there is more.

For the first time in years, icebergs have been spotted in the North Atlantic shipping lanes. The great ocean liner, the *Titanic,* struck such an iceberg on April 15, 1912, and sank within two-and-a-half hours. Not since those days have sea captains using the North Atlantic routes really been concerned with the possibility of running into these huge blocks of ice that break off from the polar ice cap and float south.

Left: Boston (January 1978) digging out of one of the
worst snowstorms to hit New England in a hundred years.
Twenty-one inches (53 cm.) of snow fell on the city over
a twenty-four-hour period. Above: ice on the Ohio River
(January 1977) threatened to halt most river transportation.

Moreover, high-altitude north polar winds have grown both in strength and in the areas of the world they affect. Coupled with a change in wind patterns, these high-altitude winds send more and more frigid air deeper and deeper south. The directional change of this polar airflow may help explain why in the winter of 1976–77, Anchorage, Alaska, enjoyed temperatures above 40°F (5°C) while Florida lost much of its orange crop to a frost. It might help explain, in fact, why the icebergs are breaking away from the polar cap.

Though their effects are not immediately evident, volcanic eruptions and earthquakes, such as the 1977 record-breaking quake in the Indian Ocean, definitely leave their mark on the world's weather. They may not, by themselves, cause an ice age, but they may very well help, as we shall see in later chapters of this book.

There are other signs—in China, the Soviet Union, Iceland, Greenland, and in many other parts of the world, including the United States—that point ominously to the coming of a new ice age. Meteorologists and other scientists are looking for these signs, evaluating and speculating on each one they detect. They talk and write about the cooling of the earth, a process that has been evident for the past fifty years and more. They are concerned about the possibly significant decrease in the amount of sunlight that reaches the earth. They speculate on the climatic effects of the increasing amount of carbon dioxide, carbon monoxide, and other pollutants in the earth's atmosphere, resulting from the increasing use of coal, oil, and other fuels in the worldwide growth of industry. They ponder the possible drastic effects of climatic change on food production. They even consider the social upheavals, rebellions, or wars that would probably accompany any dramatic climatic change, such as a new ice age.

We will explore in greater depth what these meteorologists and other scientists are thinking, their theories and speculations, in the following chapters. We will also explore their ideas on why the earth's temperature and climate may be changing.

We will also examine the glacial periods (ice ages) of the past, from the first that occurred some 250 to 300 million years ago, to the Little Ice Age that took place from about 1430 to 1850.

We will locate and examine the glaciers that are still with us.

And finally, we will explore the possible effects of a new ice age on the shape of the world itself, as well as on all the living inhabitants of this globe.

Are we on the verge of a new ice age?

Let us see.

HOW AN ICE AGE
IS BORN

Numerous ice ages have come and gone in the long history of the earth. During these periods, huge sheets of ice moved down from the polar regions, covering huge sections of the globe and destroying everything living in their path. Periodically, these massive sheets of ice receded, only to return on a time pattern we have yet to discover. We have yet to agree, too, on the precise causes for these alternating periods of glacial and "interglacial" epochs.

There are many theories, each quite logical and each perhaps correct. Accepting one theory, as we shall see, does not mean ruling out the others.

The theories that account for the coming and the departing of glacial periods fall into three main categories:

1. Astronomical theories that attribute changes in climate, and the ensuing ice age, to shifts in the earth's orbit around the sun, to the earth's rotation on its axis, or both.
2. Solar theories that claim that the sun's activities might very well precipitate the ice ages on earth.
3. Geophysical theories that contend that geophysical events—volcanic action, earthquakes, tidal waves, meteorite strikes, and any other phenomena involving interaction within our land-ocean-atmosphere system—are responsible for many changes in the earth's climate, including the ice ages.

Whatever the theory, however, there is complete agreement concerning one thing: that the temperature of the earth itself is a key factor in the movement of glaciers and the occurrence of glacial epochs.

We know that the earth is heated by the sun. We also know that there is a tremendous amount of heat that comes from within the core of the earth. The crust of the earth is only about ten-and-a-half miles (16.5 km.) deep. Below this crust is a thickish liquid mass. Inside this mass, it is believed, there may be a solid metallic ball.

The hot springs in some parts of the world and the warm spots on the top of Mount Shasta in northern California and on top of Mount Rainier in Washington (despite its crown of snow) provide evidence of the heat beneath the earth's surface. Where the flow of hot springs is blocked in some way, geysers occur. Water collects underground, getting hotter and hotter, until a build-up of pressure forces a column of boiling water to shoot up into the air. Some geysers rise only a few inches. Others reach over 100 feet (30 m.) high. Geysers are found in large numbers in Iceland, New Zealand, and also in Yellowstone National Park in the United States.

But the most dramatic and sometimes the most terrible evidence we have of the heat within the earth are the earthquakes and volcanic eruptions that often devastate huge areas of land and kill thousands of people.

The heat or energy contained in the earth is not sufficient to make the surface habitable. Without the heat from the sun, the earth would perish.

Nevertheless, the heat within the earth does affect its crust. Earthquakes split the ground, creating rivers and lakes, canyons, and mountains. The breathtaking rock formations that make up the New York-New Jersey Palisades, geologists believe, were created by such a quake. Volcanic action has moved islands up and out of the sea, where no islands were before. Other islands, such as the legendary "Lost Atlantis," have been destroyed by volcanic eruptions.

Most interesting to us in our exploration of the

Earthquakes alter the earth's crust and
wreak havoc on man-made structures.

causes for glacial movement is the theory, maintained by some scientists, that every ice age the earth has experienced has come after a period of volcanic action that built up the earth's mountains and raised its lands into broad, wide, high continents.

But volcanic eruptions themselves cannot create an ice age unless they drastically change the earth's periodic, 24-hour rotation on its axis (or its 365.242194-day rotation around the sun).

There has been no such sharp change recorded by scientists in several hundred years. But there has been a noted shift in the location of the earth's "true" magnetic poles.

In 1831, Sir James Clark Ross, the noted British explorer, after whom Ross Barrier, Ross Sea, and Ross Island were named, located the true magnetic North Pole in the Boothia Peninsula of North Canada. He placed the position of the pole at 70°51′ north latitude, 96°46′ west longitude. In 1950, it was discovered that the true North Pole had shifted to Prince of Wales Island, 73° north latitude, 100° west longitude. Another shift in the magnetic North Pole was discovered in 1965. This time the pole was placed in the southern part of Bathurst Island in Canada, at 75.5° north latitude, 100.5° west longitude. Similar shifts were reported for the location of the magnetic South Pole.

Undoubtedly a marked change in the positions of the poles would dramatically divert the sun's radiating heat from one area of the globe to another. There would be drastic changes of temperature in the different regions of the earth, and lands and seas now in the

---

The earth's true magnetic pole and two abnormal shifts that have been recorded.

[14]

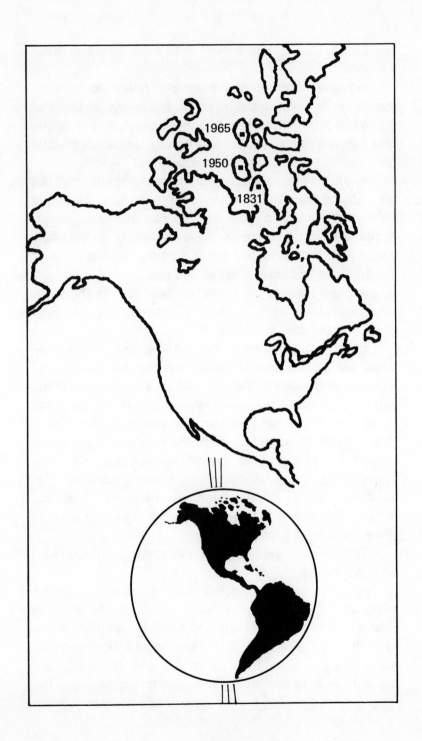

Temperate Zones might very well experience another ice age.

However, though the magnetic poles do wander, the pace of their movement is extremely slow. The magnetic North Pole has not moved out of the Arctic Sea Basin since early in the Tertiary period of prehistoric times, some 65 million years ago.

While few scientists and meteorologists dismiss the effects that the shifting of the earth's magnetic poles might have on the earth's climate, they assert rather firmly that there is no evidence of a probable major shift in the position of the poles, certainly not of a magnitude to create another ice age.

A good number of scientists believe it is the manner in which the earth orbits the sun that is more likely to be responsible for ice ages.

Early in the seventeenth century, the German astronomer and mathematician Johannes Kepler proved that the earth orbits the sun not in a circle but in an ellipse. It has also been discovered that the earth spins on its axis like a top that leans to one side. According to a number of scientists, both the rotation of the earth around the sun and its angled revolution on its axis make for cyclic variations in its space journeys. It is such variations, these scientists maintain, that are linked with the prolonged ice ages that have been visited upon the globe.

This theory, as might be expected, is not universally accepted.

All scientists are agreed that the distance of the earth from the sun, as well as the angle of the earth in relation to the sun, affects the earth's weather. Some scientists believe that this relationship between the earth and the sun (distance and angle) is largely but not entirely responsible for ice ages of the past and the

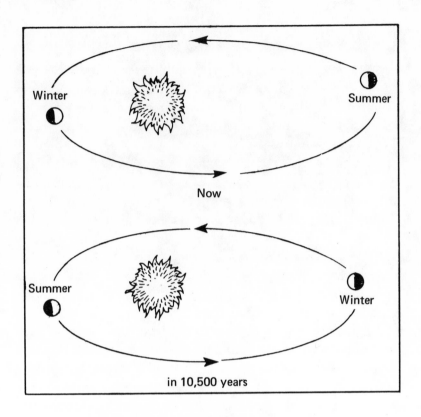

Winter

Summer

Now

Summer

Winter

in 10,500 years

The variations in the
earth's orbit around the sun.

ice age to come. Others believe that the relationship between the position of the earth and the position of the sun is responsible for only limited and short-term climatic changes.

Whatever the case, we will now examine in greater detail the relationship between the sun and the earth's weather. Perhaps we can then come more easily to some conclusions about the logic of the theories just discussed.

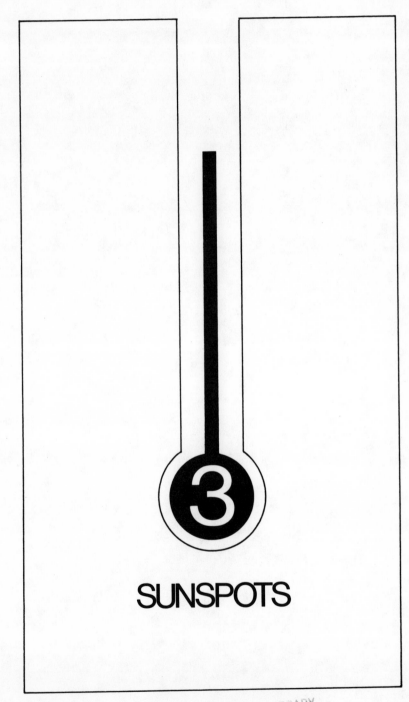

3

SUNSPOTS

It may not seem likely that the sun, an intensely hot body of gases radiating its light and heat to the earth from some 93 million miles (145 million km.) away, could be responsible for the ice ages that have come to our planet. Yet many respected meteorologists believe that the sun may very well be a prime cause of these ice epochs.

Obviously, ice ages are not caused by the direct rays of the sun, which warm the earth and supply so much of its energy, and without which we would surely die. Rather, the scientists have placed the responsibility for ice ages on what are called "sunspots."

Sunspots are groups of spots that appear, with certain regularity, on the surface of the sun. These spots are darker and 40 percent cooler than the rest of the sun's surface. They are, nevertheless, extremely hot, some 3,800° Kelvin. (Kelvin degrees are equivalent to degrees Celsius minus 273°, which would set the temperature of the sunspots at 3,527° Celsius.) In addition, these sunspots have magnetic fields at least one thousand times stronger than that of both the earth and the rest of the sun.

Some scientists believe that the sunspots are pairs of magnetic lines of force that have moved from within the sun to its surface, generally near the sun's equator. Each pair has its own polarities, positive and negative, the positive spot leading the negative across the sun's surface, till both spots fade. When these sunspots reemerge, as they do with regularity, their polarity and movement are reversed.

It seems that sunspots reach the maximum in their activity every eleven years. For this reason, certain scientists consider the sunspot phenomenon as an eleven-year cycle. Other scientists, however, consider the initial appearance and subsequent reappearance

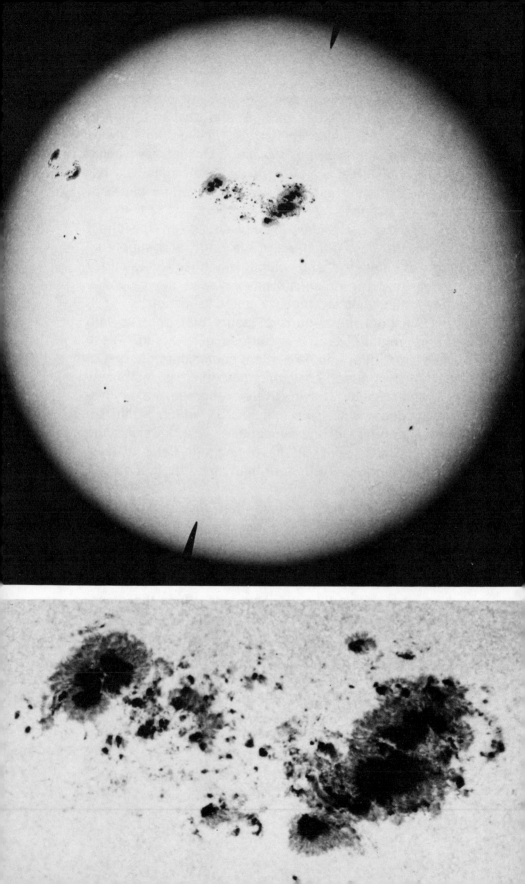

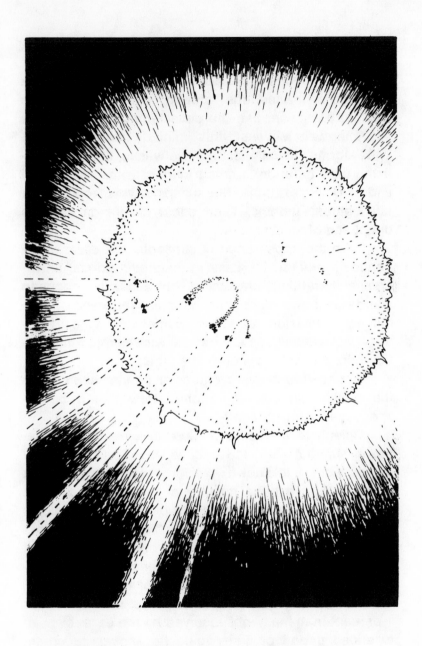

Left: the sun, and a sunspot grouping enlarged.
Above: magnetic lines of force formed from
sunspots, moving across the surface of the sun.

of the sunspots as one complete cycle, and so think in terms of a twenty-two-year cycle.

Most scientists believe that these sunspots are evidence of tremendous "storms" on the sun, storms that are somewhat like the devastating tornadoes that hit America's Midwest, but infinitely more powerful. The "storms" usually last for two weeks, though sometimes their duration is much longer. In the years 1840 and 1841, for example, the sunspots were active for a full eighteen months. Their effect on life on earth is direct and often dramatic.

With the appearance of sunspots, the earth experiences a marked increase in magnetic storms, an increase in rainfall, unusual disturbances in magnetic compasses, and more than normal interference in radio and television reception. Sometimes sunspot activity will actually knock out all wireless communication here on earth for hours at a time.

But how does this cyclic solar phenomenon affect the earth's climatic conditions? How can it be responsible for an ice age?

Scientists who believe there is a relationship between sunspots and ice epochs have quite a logical explanation. It is known, the meteorologists say, that sunspots hurl a huge number of electrically charged particles into space. A good number of these particles become trapped in the magnetic field that surrounds the earth. These particles reflect the rays or heat of the sun back into space. If there is enough such reflection of solar heat, the temperature of the earth drops. According to our scientists, a decrease of no more than 1 percent in the sunlight received by the earth over an extended period of time would be enough to create an ice epoch. However, whether or not the particles released by the sunspots and captured by the earth's atmosphere are sufficient in number and strength to

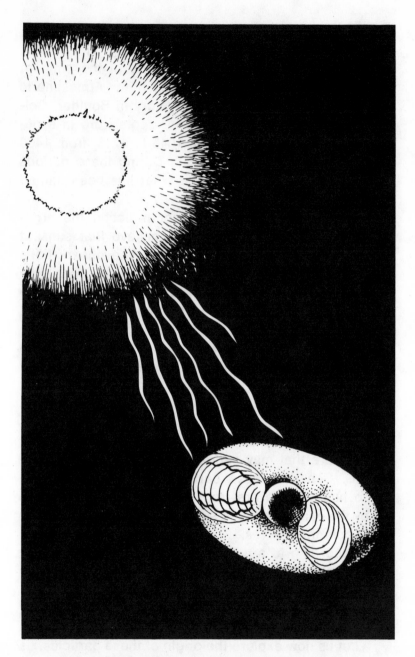

Charged particles from sunspot activity trapped
in the magnetic field surrounding the earth.

cause this 1 percent drop is still a debatable question.

Apparently contradicting this theory, and raising other questions, are the observations of the eminent John A. Eddy of the National Center for Atmospheric Research's High Altitude Observatory in Boulder, Colorado. Eddy points out that solar spots were in a period of *minimum* activity from 1645 to 1715, from 1460 to 1550, and from 1100 to 1250. During these periods of time, the earth experienced what has been named the Little Ice Age.

Eddy would *seem* to be contradicting the argument of the meteorologists who believe that sunspot activities help precipitate ice ages. The observations from Boulder, however, do not rule out the possibility of a significant relationship between the storms on the sun and the different ice epochs the earth has known.

It seems apparent that sunspots do affect the earth's climate in some way. It is either the charged particles the sunspots send into the atmosphere that cause the dramatic drop in the earth's temperatures, or it is the *lack* of such solar activity that brings those temperatures drastically down.

But there are many particles hurled into the earth's atmosphere that may seriously affect its climatic conditions. Some of these particles are minute in size, some huge. Some are from space, some are spewed up by the earth itself, and some of this debris is created by people themselves.

According to a number of prominent scientists, it is these particles, rather than those flung out from the sun, that are responsible in a large measure for the cooling of the earth, for the ice ages past, and for ice ages to come.

Let us now explore the origin of these particles, as well as the theories concerning their effect on the climatic conditions of our planet.

4

FROM OUTER SPACE

A huge amount of dust in the atmosphere, separating the sun from the earth, might well bring on an ice age. This is the theory held by a number of scientists. Huge amounts of dust may be created by the collision of an asteroid with the earth or by the crashing of an extremely large meteor or bolide (an exploding meteorite) into the earth's surface. Such sun-obliterating dust might also be created by the crash of some foreign mass into the moon. Each such collision would send an enormous amount of dust into the atmosphere. In the case of a crash on the moon, the dust from that orb would be pulled into the earth's atmosphere by the earth's gravity.

From what we have already learned about the moon, we know it has suffered many such strikes from celestial bodies. The earth, too, bears evidence of many such collisions. The earth's atmosphere protects us against the smaller objects that come our way by burning them up. But many big ones have made it through and left their marks on the surface of the globe.

The Meteor Crater, near Winslow in Arizona, is an example of the power and size of some of the meteors that have hit the earth. The Meteor Crater is nearly a mile (1.6 km.) in diameter and 600 feet (183 m.) deep. Other craters of huge dimension have been discovered near Odessa in Texas, in the Chaco of Argentina, in the northern regions of Siberia, and elsewhere. There are probably many such craters hidden by mountains or heavy foliage.

A 60- to 80-ton (52- to 72-metric-ton) meteor was found in Grootfontein, in Southwest Africa. A meteor weighing 36.5 tons (33 metric tons) was found in Greenland. A 14-ton (13-metric-ton) meteor was discovered near Portland, Oregon. Other huge meteors have been found in Brazil, Australia, and in other parts of the

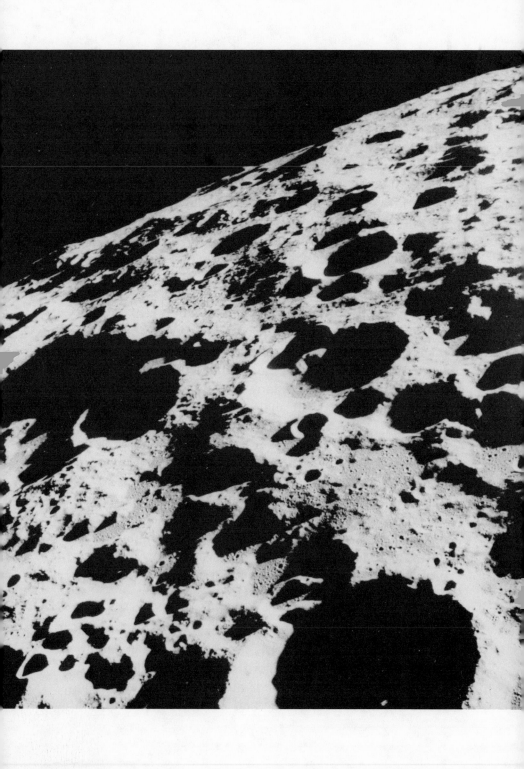

Left: an Apollo 11 view of the far side of the moon.
Note the moon's heavily pockmarked surface.
Above: the Meteor Crater near Winslow, Arizona.

world. We do not know how many more still remain to be discovered.

A meteorite rushing toward the earth moves at an incredible speed, strikes with an incredible force, gives off a tremendous amount of heat, and creates an enormous impression where it comes to a stop. All this triggers a tremendous explosion. Dust and particles of matter are hurled high and wide into the atmosphere.

The meteorite that hit the Kamchatka Peninsula in northern Siberia in 1947 was a comparatively small one, though its weight was in the tens of tons. Its impact was felt for some fifty miles (80 km.) around.

This was the most recent of the meteorites of any serious size to hit the earth. But let us keep in mind that there are some thirty thousand bodies moving around in space, any one of which might precipitously descend upon us and cause immeasurable damage.

It has been estimated that there are usually about 100,000 years between significant earth-asteroid collisions. That is a long time in relation to the life of one human being. But it is a relatively short time in the history of the earth.

A meteorite of the size that created the Meteor Crater in Arizona could wreak enormous havoc on this planet. It could send more than 4,000 cubic miles (6,300 cu. km.) of dirt and other debris into the atmosphere. It could dry up as much water and create tidal waves 4 miles (6.4 km.) high. It could produce innumerable

Two kinds of meteor rock that have struck the earth. The one above shows a heavy iron content. The one below is more stony. It weighs 745 pounds (335 kg.).

tektites (molten bodies composed largely of silica and a variety of oxides) that would pervade our atmosphere for countless years. It could even change the earth's magnetic field and create enough dust in the atmosphere to blot out for years half the light and heat we get from the sun.

It could create an ice age.

It could create an ice age in a matter of moments, according to Dr. Iben Browning, a research scientist in New Mexico, who cites as proof the finding of frozen mammoths with still unchewed food in their mouths. It would have taken a drop in temperature to −150°F (−101°C) within two seconds, he says, to freeze the mammoth so quickly. A rush of freezing wind, coming from very high altitudes, might very well create such a swift change in temperature. And such a rush of wind could result from a bolide strike. The flaming ball of fire would first hit the earth and hurl a giant mass of dirt, stones, and dust into the atmosphere. Then the mass would be pulled back to earth by the force of gravity and the result would be a devastating stream of frigid air.

A bolide striking the moon, according to Dr. Browning, could well produce a similar result on earth.

In 1937, the asteroid Hermes, about one mile (1.6 km.) in diameter, missed us by about 600,000 miles (96,5606 km.). This is not a great distance in astronomical terms. In 1947, as already noted, an asteroid hit Kamchatka Peninsula, but fortunately it wasn't large enough to seriously affect the earth's atmosphere. In 1968 we nervously followed the track of the asteroid Icarus, and it was not until the last few days, as it neared our planet, that we could be sure it would not hit us.

If Icarus had hit the earth, in addition to damaging the earth's crust and changing its magnetic field, the asteroid undoubtedly would have hurled enough dust, stones, and other debris into the atmosphere to cut off a major part of the light and heat the sun sends us.

When Krakatoa, the volcanic island in Sunda Strait, between Java and Sumatra, erupted in 1883, it sent up about one cubic mile (4.1 cu. km.) of dust 50 miles (80 km.) into the atmosphere, blotting out the sun for more than two days within a 50-mile (80-km.) radius, and for nearly a day at an observation post 130 miles (200 km.) away. This cubic mile of dust, along with its lava and ashes, was enough to diminish by 25 percent the amount of heat and light reaching the earth from the sun.

An asteroid from outer space striking the earth could conceivably send up enough dust to blot out all the radiation the sun sends us. A bolide striking the moon could conceivably do the same. Take enough sunlight and heat away from the earth and the result, inevitably, is an ice age.

Colliding asteroids and meteorites, however, are not the only events in nature that send dust and other particles into the earth's atmosphere, cutting the sun's heat from the earth and thus creating changes in its temperatures. We have already seen that a volcano, such as the one at Krakatoa, could send up enough debris to cut off 25 percent of the heat the earth receives from the sun. Let us now examine in detail these eruptive forces in nature that might well help bring about the return of glaciers to the more temperate areas of the world.

VOLCANIC ERUPTIONS

For a long time it has been recognized that volcanic eruptions have had a marked effect on climatic conditions. The giant quantities of steam spewed up by a volcano condense, rise, and then descend upon the earth in the form of torrential rains. It has also been widely observed that a decline in the earth's temperature always takes place in the area of a volcano following its eruption.

Recent studies have developed more dramatic proof of the relationship between volcanic activity and the earth's climate.

Some 30,000 years ago, the temperature of the Antarctic regions declined between two and three degrees Fahrenheit (about one degree Celsius) during a time of extensive volcanic activity in the area.

Between 16,000 and 17,000 years ago, similar volcanic action brought the earth its lowest temperatures ever.

As we have already noted, volcanoes hurl up tremendous amounts of dust. In the opinion of many meteorologists, a marked increase in the quantity of dust that normally exists in the atmosphere could lower the temperature of the earth six to ten degrees Fahrenheit (three to five degrees Celsius). Such a decrease in the earth's temperature, if it lasted for even a few years, could very easily result in the development of an ice age.

At one time it was believed that the effects of volcanic eruptions on the earth's climate were only temporary. Whatever climatic change resulted from volcanic action, however intense, would be of short duration. We now know that the eruption of volcanoes, if they are sufficient in number, strong enough, and persist over a considerable period of time, can bring about dramatic changes in the earth's climatic conditions.

A nighttime view of the eruption of
Mt. Vesuvius, Italy, in 1944. Note the
molten lava snaking its way downward.

There are, generally speaking, four main kinds of volcanic eruptions: Hawaiian, Strombolian, Vulcanian, and Pelean.

The Hawaiian eruption sends out quantities of basalt, a hard dense rock. There is no explosion, as there is with other volcanic eruptions. Mauna Loa, in the Hawaiian Islands, is an example of such a volcano.

The Strombolian eruption, named after the volcano Stromboli in the Aeolian Islands (or Lipari Islands) north of Sicily in the Mediterranean Sea, sends out a continuous flow of lava during intermittent mild explosions.

Vulcanian eruptions are more explosive. A mass of molten matter, which eventually becomes lava, accumulates just beneath the opening of the volcano. Here it stays, blocked by a mass of hardened lava. Gases unable to escape through the hardened lava build up pressure. When the hardened lava can no longer hold back the mounting pressure, the gases explode and hurl both solid and liquid rock into the air, plus clouds of steam and dust.

The Pelean eruption, named after the volcano Pelée in Martinique in the West Indies, is the most violent of volcanic eruptions. Pelée is known to have erupted in 1792 and 1851, but on May 8, 1902, its eruption was so violent that it destroyed the city of Saint Pierre, killing 40,000 people.

A Pelean eruption spews forth, in addition to hard and molten rock, a hot, heavy cloud of mixed gases and glowing particles that moves at the rate of 100 miles (161 km.) an hour and destroys every living thing in its path.

There have been many other such volcanic upheavals. The volcanic action that brought the temperature of the Antarctic down two or three degrees Fahrenheit (one degree Celsius) some 30,000 years ago was one. So were the eruptions of sixteen and seventeen thousand years ago.

In comparatively more modern times, the eruptions of Mount Vesuvius in Italy in 63 A.D. and again in 79 A.D. killed more than 2,000 people and buried the cities of Pompeii and Herculaneum. Vesuvius erupted again at irregular periods until the year 1138, and once more in 1631. This last time, aided by earthquake and tidal wave, it killed 4,000 people.

Mount Etna in Catania, Italy, erupted in 1669, killing more than 20,000 people. In 1783, Mount Laki in Iceland erupted and killed one-fifth of the population.

The volcanic eruption of Mount Tamboro in Indonesia in 1815, aided by tidal waves, took the lives of some 12,000 people. Krakatoa, mentioned earlier, destroyed two-thirds of the island on which it stood and killed almost 40,000 people. In 1963, Mount Agung erupted on the idyllic island of Bali, killing at least 15,000 women, men, and children.

Volcanoes erupt in every part of the world. There are some 600 to 800 volcanoes that are known to be active, sending out a steady or intermittent stream of lava, or that are on the verge of exploding. One belt of these volcanoes stretches around the world from the Aleutian Islands (the tip of the Alaskan peninsula) through the Pacific coasts of North, Central, and South America, Japan, the Malayan islands, and the South Seas to New Zealand. Another belt of volcanoes reaches from Central America, through the West Indies, the Atlantic Ocean, the Mediterranean Sea, and Asia Minor to the East Indies. Volcanic action takes place on

An aerial view of the crater of an active volcano on the island of Oshima, Japan.

the bottom of seas and oceans, as well. It is such action that swallows islands and gives birth to new land-masses.

There has been no way devised yet that can alert us to an approaching volcanic catastrophe. Mount Shasta and Mount Rainier have warm spots on their surfaces, but that is no indication of an imminent eruption. Occasionally there are rumblings in the craters of volcanoes, but the rumblings may last for years and nothing come of them. On the other hand, people have visited and even picnicked in craters of seemingly "dead" volcanoes only days, or even hours, before they violently erupted. We know that gases build up in the interior of these volcanoes, then explode through the plug that had kept them imprisoned.

We cannot tell when such explosions will occur. Nor are we certain about what causes these gases to build up within the volcano, though there are a few theories. One theory has it that the gases result from the excess energy the earth receives from a heavy meteorite strike. Another theory is that tidal waves are in a way responsible for volcanic activity. It is not that tidal waves *cause* the volcanic eruption, but rather that they *trigger* it. It has already been noted that two violent volcanic eruptions, that of Mount Vesuvius in 1631 and that of Mount Tamboro in Indonesia were accompanied by tidal waves. Whether the tidal waves were caused by the eruptions or the eruptions were triggered by tidal waves is yet to be decided.

We do know that volcanic eruptions hurl huge amounts of dust into the upper atmosphere. The dust lingers in the atmosphere and spreads all around the earth, definitely interfering with the amount of radiation the earth gets from the sun.

Dr. Charles G. Abbott, the renowned astrophysicist associated with the Smithsonian Institution, noted a marked decrease in the solar radiation received by the earth in 1885, 1891, and 1892. He attributed this decrease to the volcanic action of the time.

A study of the effects of the eruption of Mount Katmai in Alaska in 1912 produced a similar report from astrophysicists. These scientists, reporting from various observation posts, said that they found a decrease of as much as 20 percent in the radiation the earth received from the sun that year. They attributed the decrease to the haze created by the volcanic dust erupting from Mount Katmai.

This dust, in addition to the hot gases spewed out of the volcano's crater, contains a myriad of fine particles that penetrate the earth's atmosphere and move out into the stratosphere. Krakatoa hurled out about one-and-a-quarter cubic miles (5.2 cu. km.) of such dust. Though other volcanoes have thrown up considerably smaller quantities of this dust, still others have spewed up to four times as much, and more.

Much of this dust is sulphur dioxide. Particles of lava emit sulphur dioxide at high altitudes. Ultraviolet radiation breaks up sulphur dioxide into its component parts, sulphur and oxygen, but this action takes place primarily in the region above the earth's ozone layer. Beneath this ozone layer the hot lava evaporates and condenses into a gaseous suspension, aerosols in effect.

The effects of aerosols on the temperature of the earth has aroused much debate in recent years among scientists and in the general public, but without any firm conclusions. Whatever the outcome of the controversy, however, we know that sulphate particles do reflect the light of the sun back into space. And we know that clouds of such sulphate particles, created by volcanic eruption, decrease the earth's temperature by a considerable number of degrees.

[45]

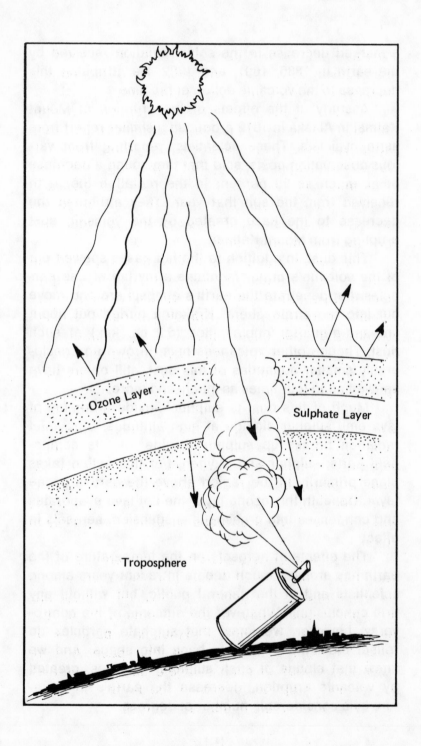

Ozone Layer

Sulphate Layer

Troposphere

Carbon dioxide particles, too, are present in these clouds, and carbon dioxide has a warming effect on the earth's atmosphere. This effect has been termed the "greenhouse" effect, which we shall soon explore in depth. The sulphate particles, however, largely negate the effect of the carbon dioxide and the end result is a colder climate anyway.

And, as has already been indicated, if there is enough volcanic action within a limited span of time and enough sulphur dioxide-laden dust spewed out through craters to cut down the amount of radiation we receive from the sun, then this planet will experience, as it has experienced in the past, the coming of still another ice age.

The aerosol effect.

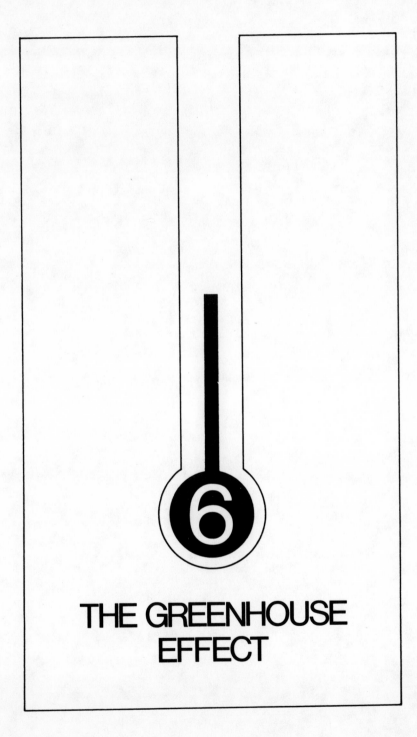

6

THE GREENHOUSE
EFFECT

Meteorological records indicate that the earth experienced a worldwide warming period between 1800 and 1940. These records also indicate that the average worldwide temperatures of the earth from the 1930s to the 1950s were at their highest levels in several thousand years. Scientists think they know why this happened.

Mark Twain is reported to have remarked on several occasions that everybody talks about the weather but nobody does anything about it.

Not true, as we shall see. People have done a great deal about the weather, and continue to do so. At least it would seem so, if we examine what was going on with the human race between the years 1800 and 1940.

It has already been mentioned that carbon dioxide in the atmosphere tends to warm the earth. Carbon dioxide is a gaseous compound of carbon and oxygen. Shortwave radiation that comes from the sun can pass right through it. It is less easily penetrated by the long-wave radiation that is emitted by the earth's land and water, as it is heated by the sun.

Carbon dioxide in the air, then, does nothing to obstruct the heat the sun sends us, but it prevents from escaping into the stratosphere a considerable amount of heat from the earth. Consequently, the more carbon dioxide in the earth's atmosphere, the higher will be its temperature. This is the so-called "greenhouse effect."

A greenhouse, or hothouse, is a glass-enclosed structure in which plants, flowers, vegetables, and even some fruit are kept warm so that they may survive the colder weather or be grown out of season in artificially warm temperatures. The warming of the earth by carbon dioxide is, in a sense, an artificial process, too.

This artificial process is our doing. More correctly, it is the doing of the machines we have created. It was in the middle of the eighteenth century that people and their machines began to send enormous and ever-increasing amounts of carbon dioxide into the earth's atmosphere.

The middle of the eighteenth century is the date generally given as the beginning of the Industrial Revolution. It was at this time that society began to change from an agricultural to an industrial way of life. Industry required, at the beginning, the use of coal in great quantities. Later it would use gas, oil, and nuclear energy. All these fuels produce carbon dioxide.

We rarely talk of the Industrial Revolution today, except in history and economics classes. We speak of the Atomic Age in which we live. Still, the effects of the Industrial Revolution that swept the world of our great-grandparents are still measurably with us.

There are some 130 million cars, trucks, buses, and jet aircraft in the United States alone, consuming millions and millions of gallons (liters) of fuel. Factories, power plants, home heating units, incinerators, and so on consume enormous amounts of coal and oil. It has been estimated that this burning of coal, oil, and natural gas sends at least six million tons of carbon dioxide a year, along with other waste, into the earth's atmosphere.

The United States is the most industrialized nation in the world, but Great Britain, France, Germany, the Soviet Union, and Japan, among others, are scarcely less active, spewing their tons of carbon dioxide into the air that surrounds us.

In 1952, a "pollution shroud" was responsible for the death of some 4,000 people in London, England. A similar "shroud" killed 20 people in Donora, Pennsylvania, in 1948. In Yokkaichi, Japan, because of the polluted air, schoolchildren wear gas masks when they play outdoors.

Not all these pollutants are carbon dioxide, but there was, or is, enough carbon dioxide in the earth's atmosphere to affect its temperature. Certainly it was the enormous quantity of carbon dioxide, spewed into the atmosphere by the vastly industrialized world, that was largely if not entirely responsible for the unusually high temperatures of the earth in the years from 1800 to 1940.

The earth has cooled off considerably since 1940, but not enough for those scientists who believe the greenhouse effect is still potentially dangerous. It is believed by these scientists that if people and industry continue to consume coal, gas, oil, and nuclear energy at the current rate (actually we are using these fuels at *accelerating* rates), we may well be faced with disaster. The increase in temperatures brought on by the increase of carbon dioxide in the earth's atmosphere, they argue, could melt the ice in the Antarctic regions in no more than twenty, thirty, perhaps forty years. The melting of the Antarctic ice, they say, would raise sea levels all around the world. Tokyo, New York, and most of the other great cities of the world would be submerged under water. So would much of the United States. Farms, industrial plants, and power plants would be destroyed. Millions and millions of people would be forced on a desperate search for dry land.

On the other hand, there are as many scientists who believe that all the smoke and dust and all the other particles hurled into space over the years by our car exhausts, our factory chimneys, and our power plants have acted to screen out much of the radiation of the sun. They say that the debris sent up into the atmosphere by humankind and its machines has nullified the greenhouse effect. They point out that even though

[53]

there is 10 to 15 percent more carbon dioxide in the air today than there was a hundred years ago, the temperature of the earth has gone down by almost three degrees Fahrenheit (one degree Celsius) since 1950, and continues to fall.

Actually, the Northern Hemisphere has cooled off by less than 1 percent. But in the more frigid regions, the drop in temperature has been more marked. In Iceland, the readings on the thermometer have dropped somewhere between four and five degrees Fahrenheit (two degrees Celsius). Dr. R. S. Bradley and Dr. G. H. Miller, scientists from the University of Colorado, have reported an increase of snow and ice on Baffin Island, north of Hudson Bay. They have also noted that the old glaciers on the island have grown bigger, and that two new glaciers have been born.

It is the sharper drop of temperature in the colder areas of the earth, the growth of old glaciers, and the birth of new ones that interest us more at the moment. We will investigate these phenomena as possible signs of the coming of a new ice age.

7

GLACIERS

When the temperature of the earth's atmosphere drops low enough, great sheets of ice at the poles begin to move toward the more temperate zones of the globe. The Antarctic region consists of almost 5 million square miles (about 13 million sq. km.) of ice in the form of glaciers. The Arctic region stretches almost a third of the way from the pole to the equator. We are learning more and more about the surface topography of these ice fields. We do not yet know the depth or the shape of what lies beneath this surface. We do know that there is movement in these masses of ice, the glaciers.

Glaciers are formed by the compacting of snow. At some time you have probably noticed how a snowball became a solid ball of ice as you compacted the snow in your hand. As these glaciers grow in size, the sheer pressure of their dimensions sets them into motion. Glaciers of such size are found not only at the poles but also in high mountains, high plains, and high plateaus throughout the world.

There are valley or mountain glaciers that originated in the mountains and were sent down through the valleys, eventually to be contained by valley walls. There are Piedmont glaciers that formed at the foot of mountains, as valley glaciers spread and joined together. There are glaciers that are called ice caps. These cover entirely, or "cap," ranges of mountains and valleys alike. There are the continental glaciers. These, as their name implies, cover areas as large as continents.

There are some 1,200 valley glaciers in the Alps alone. The continental glaciers, which would cover much of the earth during a new ice age, are presently confined to Greenland and Antarctica.

Glaciers, whatever the year or season, are never completely still. Their movement forward, backward,

or sideward is constant. They move as solid masses of ice, and they take the direction of least resistance. The rate at which they move depends on their volume and, of course, on the nature of the land and water over which they are moving. If the forward edge of the glacier is melting faster than the rate at which the glacier is moving, then the glacier will retreat. Conversely, if the glacier's forward edge is melting at a slower rate, the glacier will continue its forward movement.

During the summer, Alpine glaciers are known to move from ten to twenty inches (25.4 to 50.8 cm.) in a single day. The edge of the continental glacier covering Greenland, on the other hand, will move just a few inches (cms.) a week in the summertime.

Glaciers do not all move in unison in one direction, forward or backward, at any given time in their history. Observations of ice caps in Alaska, Scandinavia, and Patagonia in South America, show us that some glaciers will move forward at the same time that others in the same area move back.

During the creation of an ice age, however, the glaciers do move consistently in the same direction—forward. And as they go they alter the shape of the land. They create great crevasses in mountains and fjords (as in Scandinavia). They cut down mountains and broaden valleys. They cover lakes and rivers with masses of ice. They also destroy whole forests, everything that grows in their path, and entire species of animal life. A new ice age would obliterate the human efforts of perhaps two thousand years.

The *Mer de Glace* in France.
This is one of Europe's largest
glaciers. As its French name
indicates, it is truly a "river of ice."

8

ICE AGES PAST

At one time it was believed that the earth experienced just one prolonged ice age, during the Pleistocene epoch, a million years ago. In 1837, Louis Agassiz, the Swiss-American geologist, formally presented this theory to the world. Since then, however, further investigations into the geological history of our planet, inspired by Louis Agassiz himself, have cast considerable doubt on his concept.

First, we realize that we cannot give precise dates for events that occurred so long ago. Some scientists place the beginning of the Pleistocene epoch at a date closer to our time, about 800,000 years ago. The majority of scientists, however, believe that the Pleistocene epoch and the Ice Age began between 2 to 3 million years ago, though a good number of geologists say that both had their beginnings even earlier in time.

Second, the theory that the Ice Age was limited to one prolonged period has also been more or less discarded. There is evidence in some glacial-like deposits and in some recently uncovered erosion, for example, that points to the possibility that the earth experienced an ice age as much as 700 million years ago, and perhaps others even earlier in its history.

Most of what we know about ice ages, however, is limited to our knowledge of the Ice Age that occurred during the Pleistocene epoch.

We know, for example, that this epoch witnessed not one but four principal periods of glaciation. These periods, observed first in the Alpine region, were called "Wurm," "Riss," "Mindel," and "Gunz," after streams in the Alps, where valley glacial deposits were of distinguishably different ages. Somewhat later, geologists working in America discovered similar differing layers of glacial deposits, and they called the four glacial periods "Wisconsin," "Illinoian," "Kansan," and "Nebraskan."

Some effort was made by geologists to match up the Alpine with the American periods of glaciation, as to time of occurrence. This effort has not succeeded. Still, we are fairly certain that the earth did experience four periods of glaciation during the Pleistocene epoch. That is, during the approximately 1.5 million years of that epoch, there were four periods during which the glaciers advanced, retreated, and advanced again to cover great areas of the earth. At the peak of the Pleistocene epoch, sheets of ice as thick as two feet (.6 m.) covered much of North and South America, Europe, and Asia.

In Europe, much of northern Germany and the Russian and Polish plains were buried by the glaciers. The ice stretched from the Scandinavian countries almost to the Crimea in the Black Sea. It covered most of England and all of Scotland and Ireland. Switzerland, too, lay under ice, as did much of France. The Pyreneean Mountains, which separate France from Spain, lay under glaciers, as well. So did the Sierra Nevada Mountains, at the edge of the Mediterranean Sea.

In North America, Greenland was completely covered by ice. So was Canada. In the United States, the glaciers blanketed an area extending from New Jersey and Pennsylvania to the Ohio and Missouri Rivers, and into North Dakota, Montana, Idaho, and (the state of) Washington. In South America, Patagonia and the southern Andes lay under the glaciers from the Antarctic. In Asia, the glaciers moved down far enough south to cover the Himalaya and the Caucasus mountain ranges.

At least four species of elephants in North America were destroyed by this Ice Age, two of these species larger than the modern elephant. The mammoths were destroyed. Though some mastodons seemed to have survived, the species, for all purposes, was eliminated from the North American continent. Ten species of horses, most the size of ponies, were also killed off. Camels, giant beavers, woodland muskoxen, large bison, and a variety of cats, among other animals, also disappeared from the North American continent during the Pleistocene epoch. Wild pigs, for example, which lived in the northern areas of America, are now found only in Texas, Mexico, and Central America.

There were other effects, of a somewhat more pleasant nature, caused by the Ice Age, and these might well be noted too.

High pressure from the Arctic region pushed the normal Atlantic rainstorms southward. The cyclonic winds, common to central Europe today, moved over the Mediterranean, through the northern areas of the Sahara, across the Tigres and Euphrates rivers into what is now Yemen and Saudi Arabia, and through Iran and India.

The Sahara Desert was not a desert at all during the Pleistocene epoch. It was a green and thriving place. So were the presently parched lands of the Iranians, and the Indians.

At a time when only the wooly rhinoceros, the mammoth, and the reindeer could survive in the south of England and France, there was animal and vegetable life in North Africa that can only be found today in the central and southern areas of that huge continent.

Some scientists believe that it was during this Pleistocene epoch, as well, that humans, as a distinct species, began to emerge as *Homo erectus* (erect man). The *Homo erectus* was to develop, it is believed too, into *Homo sapiens* (wise man), men and women not markedly different from men and women of today. It was during this same Pleistocene epoch, it is further

[65]

believed, that humankind became meat-eaters and hunters. This is all theory, of course, and the theory has been disputed by some anthropologists who claim that some forerunners of humankind walked erect and were possibly meat-eaters before the Pleistocene epoch.

Some 20,000 years ago, according to most meteorologists and geologists, during the last part of the Pleistocene epoch, the period known as the Wisconsin stage, the glaciers showed signs that the Ice Age was truly waning. There has been no glacial movement of its proportions since. However, the threat of a new ice age has never left the earth. It has even been suggested by a number of geologists that we are still in the Wisconsin stage. They say that it is still waning but that every once in a while it gives evidence that it is still very much alive, and perhaps still dangerous.

Between 1000 B.C. and 1 A.D., after an apparently long period of inactivity, the glaciers began to move again. Geologists termed this movement "neo-glaciation," or new glaciation, but it might very well have been a sign that the Wisconsin stage had not yet entirely left us.

In the thirteenth century, there was greater movement of the Greenland glacier. It did not move down far enough south to threaten life in southern Canada, the United States, and Europe, but the increased movement lasted for more than five hundred years, reaching its maximum limits about 1750, before it began to retreat. But its retreat was slow, and there was still evidence of the glacier's presence below the pole as late as the last years of the nineteenth century.

This particular movement of the Greenland glacier has been termed the Little Ice Age. It may have been "little," but it was big enough to empty Greenland of the Norsemen who had settled there.

In the Alps, about seven hundred years ago, even the comparatively minor movements of its glaciers were enough to destroy whole forests.

It is believed by some respected geologists and other scientists that there is a cycle to glacial movement. These scientists have done their research well, drawing graphs and tables to prove that glacial thrusts occur every 854 years. However, this is rather a too precise figure for scientific prophecy, and there are as many eminent scientists who say that it takes several thousand years for an ice age to occur.

The last ice age, discounting the Little Ice Age that some geologists claim was still part of the Wisconsin stage, did occur only a few thousand years ago. And if the 854-year-cycle theory is correct, then the earth is certainly due, overdue, for another glacial thrust. And a glacial thrust may very well develop, under conditions beneficial to such thrusts, into a full-fledged ice age.

Now let us examine the climatic conditions on the earth today to see whether or not they could produce such glacial thrusts and the coming of another ice age on earth.

9

SIGNS OF THE
TIMES

Robert Dickson, who is a member of the Long Range Prediction Group of the United States Weather Service, has recently called attention to the unusual activity of the polar high-altitude winds. It is these winds that gave the eastern half of the United States its unusually heavy snowstorms and cold weather in the winter of 1976–77.

The question is, will this unusual pattern, which brings freezing temperatures to Florida and the states bordering the Gulf of Mexico, become normal?

Dr. Reid A. Bryson, director of the Institute for Environmental Studies at the University of Wisconsin, says that the Northern Hemisphere has been cooling steadily since about 1950 and that the Atlantic Ocean, during this period, has cooled one degree Celsius.

Dr. Walter Orr Roberts, an esteemed scientist, is of the opinion that this downturn in temperature, which began in the 1950s and continued more rapidly in the 1960s, was not accidental, and is not likely to be of short-term duration. He sees a continuing drop in the earth's temperature for another twenty or thirty years. Such a continued, persistent drop in temperature might well trigger the development of another ice age.

According to a number of meteorologists, the earth is already experiencing the markedly colder weather that prevailed in the years 1600 to 1850, during which time the earth was in the Little Ice Age. A number of meteorologists believe that if this cooling trend continues long enough, the earth will suffer a new ice age that might well cover large parts of Europe, the Soviet Union, and Canada, as well as the entire northern third of the United States.

What is the evidence to support these opinions and theories? For one thing, there has definitely been a drop in the earth's average temperature over the past thirty years, as professional observers have carefully and scientifically noted. The temperature has dropped almost a full degree Fahrenheit (one-half degree Celsius) in the Temperate Zones. It has dropped almost five degrees Fahrenheit (two-and-one-half degrees Celsius) in Iceland, Greenland, and the polar regions. The average drop in temperature for the entire earth is not quite three degrees Fahrenheit (two degrees Celsius) but, significantly, that average continues to fall.

The quantity of ice in the polar seas, as a result of this fall in temperature, is greater than it has been for perhaps hundreds of years.

In the Canadian Arctic, sections of Baffin Island where there used to be some green in the summertime are now completely covered with snow throughout the entire year.

United States weather satellites, which roam over the Northern Hemisphere, report an increase of at least 12 percent in the permanent snow and ice cap covering the polar regions.

And the largest glacier in mainland Canada has begun to advance at a rapid rate, as compared with the normal rate of glacial movement.

We mentioned earlier that icebergs, those great blocks of ice that break off from glacial masses, have returned in considerable number to the North Atlantic shipping lanes.

The movement of warm-weather animals southward has been noted and this is occurring in the more *temperate* climates of the Temperate Zones.

In the Soviet Union, because of the threat of approaching glaciers, a considerable number of people have been evacuated to considerably safer places south.

To return to opinion for the moment, Dr. J. Murray Mitchell, senior research climatologist for the National

Oceanic Atmospheric Administration, has warned that, from the point of view of agricultural productivity, there is little prospect of an improving climate for the earth. On the contrary, he believes that it will only get worse.

We have already lost the use of certain food-growing regions in Canada, the Soviet Union, and northern Europe, due to the descending temperatures in these areas. India and China have suffered decreasing harvests because shifting polar winds have been changing rain patterns all over the world.

"We are going to have to brace ourselves for the prospects of a lot of poor harvests," says Dr. Mitchell.

The pattern seems to be clear. Certainly there is enough informed opinion, theory, and evidence to indicate that a new ice age is more than a possibility. It is a distinct probability.

The extent of the coming ice age is more difficult to estimate. Will the huge glaciers stop at the northern edges of Canada, Siberia, and the Scandinavian countries, or will they move down to cover the British Isles, much of France, Germany, the Soviet Union, and fully a third of the United States?

Of course we must remember that glaciers move extremely slowly. It may take hundreds of years before an ice age reaches its peak strength.

In any event, let us examine the possible effect on the world, its geography, its animals, its vegetation, and its people to see what will happen if, indeed, another ice age is coming.

10

A CHANGED WORLD

An ice age of comparatively minor dimensions, like the Little Ice Age that peaked about 1750, would create tremendous problems for the people of the world. The greatest problem would be food: how to prevent mass starvation.

An ice age like that of the Kansan or Wisconsin stage of the Pleistocene epoch, huge in dimensions and thousands of years in duration, would bring about dramatic changes in the face of the earth and calamity to its people.

The higher latitudes of the Temperate Zones would be covered completely by the massive glaciers. Much of the Appalachian and Rocky Mountains, with all their running streams of water, would be buried under the ice. So would the five Great Lakes, the lochs of Scotland, and the numerous lakes in Canada. The St. Lawrence River, the Hudson, the Colorado, the Columbia, and much of the Mississippi would lie frozen under the huge glaciers. So would the great wheatlands of Canada, China, the Soviet Union, and the United States.

On the other hand, shifting winds would undoubtedly bring rain and moisture to the deserts around the Black and Caspian seas. The Gobi Desert and the Sahara would most likely become fertile land again. The Mojave Desert and Death Valley, in the southwestern part of the United States would grow green.

There would be no vegetation to keep the moose, the grizzly bear, and the Alaskan kodiak alive in their present environs, and they would move south. Except for the penguin, there would be no birds where the ice destroyed arable lands. The Great Banks, now so rich with fish, would be gone, perhaps forcing even the penguins to leave their habitat.

Even the slightest drop in the temperature of the earth drastically reduces the harvesting of grain around the world and brings famine and starvation to millions. Kenneth Hare, University of Toronto climatologist, speaking of the failure of the Russian harvest in 1972, said, "I don't believe that the world's present population is sustainable, if there are any more than three such crop failures in a row."

An ice age, however limited in size and duration, would bring three times three such crop failures, and more. During the ice ages of the Pleistocene epoch there were no harvests at all in areas where grain now grows in relative abundance.

If there is no grain, people, like birds and animals, will move to where they can find wheat, fruit, and meat to sustain themselves. This has happened in the past many times since the beginning of human history. It will happen again should an ice age descend upon us.

In primitive times, tribes moved from place to place in their search for food. If another tribe occupied the new land, the wanderers fought and killed to take it over.

Between four and five thousand years ago, during the Early Bronze Age, there was something of a population explosion. Whole villages and towns found it impossible to reap enough food to feed themselves. There were great migrations, great battles, and much bloodshed over who was to get the good arable lands still available.

The Sumerians were, to the best of our knowledge, the first people to settle the Fertile Crescent, the area in Ancient Mesopotamia around the Mediterranean, so-named because of the abundance of food it produced. But Semitic tribes, about 4,000 years ago, their eyes and their appetites on the Crescent, moved in and threw the Sumerians out of their homeland.

As a matter of fact, there were a number of different Semitic tribes who fought for the food of the

Fertile Crescent, and their fighting lasted over a period of some four hundred years.

The Exodus in the Bible tells the story of the wanderings and battles of the Hebrew people in their search for the promised "land of milk and honey."

Much more recently, at the time of the Little Ice Age, Eskimos moved down from the north and swept into and attacked the small Norse settlements in Greenland in search of food.

In modern times, the earth has been plagued by wars, great and small, struggles between nations for "living room," another name for land that produces rye and wheat and corn and barley.

Obviously, an ice age, with its destruction of so much arable land in the Temperate Zones, would serve only to intensify the struggles and wars among an increasingly hungry people.

Could we expect the migrations from the northern regions, in the event of a major ice age, to be peaceful migrations? Would Americans welcome the displaced and hungry Canadians should the glaciers cover Canada?

And if a new ice age should send sheets of ice deep into America, destroying cities and plains, certainly Americans would have to look south, to Mexico, for living room and for food. Would the move into Mexico be peaceful or would the Mexicans resist? Would people find it necessary to turn to war for the resolution of the problems brought on by a new ice age?

Americans have always been at peace with the Canadians and nearly always at peace with the Mexicans. But it is not likely that, with millions of people made homeless by the descending sheets of ice from the north, peace would prevail among the peoples of North America.

In Europe, at one time or another, the Swedes

have invaded Russia, Russia has invaded Poland, and Germany has invaded France three separate times in the past hundred years. The political complications, in the event of even a minor new ice age, would prove infinitely more complex in Europe than it would in the Americas. The need for food and shelter and the forced migrations would undoubtedly make for the bloodiest conflicts on that continent, which would eventually spill over into Asia.

Would China and India, in that largest of continents, welcome hungry, land-seeking refugees from Russian Siberia?

A new ice age, even the threat of a new ice age, it seems evident, would create a catastrophic situation for the tremendous number of people who inhabit the more northern regions of the earth. It would bring about a devastating insufficiency of food. It would develop a frantic and hungry search for safe and arable land. It would incite wars, with nation battling against nation for sheer survival.

It is not a pretty picture.

How far or near, in time, we are from such a catastrophe cannot be predicted with any precision. But there are ominous signs coming from the polar regions and showing up in our atmosphere.

Is there anything we can do to prepare ourselves for this eventuality?

Dr. Stephen H. Schneider, deputy chief of the Climate Project at the National Center for Atmospheric Research in Boulder, Colorado, urges that we store the extra food we reap in the good years so that we do not go hungry in the years of poor harvest.

Undoubtedly this is good advice, and undoubtedly we should follow it. But how much grain can we store? And could we ever store enough to withstand the onslaught of a severe and life-killing new ice age?

SUMMING UP

The evidence presented by meteorologists, climatologists, and other scientists indicates that the earth is on the verge of significant change. We cannot say with any certainty whether this change will be one more phase of the ebbing Wisconsin stage or the development of a new ice age. Nor are we certain on how far south the moving polar glaciers will travel.

We know that the earth has experienced a number of major ice ages and what is called the Little Ice Age in its long history.

We know that there has been a change in the direction of the high-altitude polar winds, that the Canadian glaciers are moving south at an unusually rapid rate, and that the number of icebergs in the North Atlantic shipping lanes has increased considerably.

We know that the temperature of the earth has dropped an average of 4 to 5 percent in the past thirty years, to temperatures that prevailed during the Little Ice Age, and that a continuing drop in temperature these next twenty or thirty years has been predicted.

We have been informed, as well, that a 1 percent decrease in the amount of sunlight received by the earth would cool the planet sufficiently to create an ice age in a matter of no more than a few hundred years. The carbon dioxide, dust, and other debris produced by volcanic eruptions, meteorite strikes, bolides, and by people themselves, may have been at one time responsible for the greenhouse effect that warmed the earth. But, in the years to come, that same debris might just as easily be responsible for diminishing the radiation the earth receives from the sun, and by more than that significant 1 percent.

However, as one person's life is measured, a few hundred years is a long time. Nature may go through many dramatic changes in those few hundred years. It would be good to think that with all our increasing scientific knowledge, we might be able to stave off the catastrophe of an ice age. Realistically, however, nature seems to have a mind and way of its own.

At a national convention of scientists in Denver, Colorado, in February 1977, the astronomer Dr. John A. Eddy said that the possibility of a new ice age is at least thousands of years away. This he said was the "most pessimistic" view.

Nevertheless, though we cannot predict their dimensions, nor how far south they will travel, we do know that the polar glaciers are on the move.

For all purposes then, and whatever else we may think or believe, we may be fairly certain that in the course of natural events, a new ice age cometh.

BIBLIOGRAPHY

Bibby, Geoffrey. *Four Thousand Years Ago.* New York: Knopf, 1962.

Brooks, C. E. P. *Climate Through the Ages.* New York: Dover, 1970.

Chase, Allan. *The Legacy of Malthus.* New York: Knopf, 1976.

Childe, V. G. *Prehistoric Migrations in Europe.* Cambridge: Harvard University Press, 1951.

Clairborne, Robert. *Climate, Man and History.* New York: Norton, 1970.

Cornwall, Ian. *Ice Ages: Their Nature and Effect.* London: John Baker, 1970.

Daly, Reginald A. *The Changing World of the Ice Age.* New York: Hafner, 1924.

Davis, Darrell H. *The Earth and Man.* New York: Macmillan, 1944.

Flint, Richard F. *Earth and its History.* New York: Norton, 1973.

Frankfort, Henri. *The Birth of Civilization in the Near East.* Bloomington: Indiana University Press, 1951.

Gates, David M. *Man and his Environment.* New York: Harper & Row, 1972.

Grippen, J. *Our Changing Climate.* London: Faber, 1975.

Huntington, Ellsworth. *Civilization and Climate.* New Haven: Yale Press, 1942.

John, D. S. *The Ice Age: Past and Present.* Glasgow: Collins, 1977.

Lantier, Raymond. *Man Before History.* New York: Walker, 1965.

Lockwood, J. *Climatology: An Environmental Approach.* London: Edward Arnold, 1974.

Orlinsky, Harry M. *Ancient Israel.* Ithaca: Cornell University Press, 1954.

Schuchert, C., and Dunbar, Carl O. *Outline of Historical Geology.* New York: John Wiley and Sons, 1944.

Schultz, Gwen. *Ice Age Lost.* Garden City, N.Y.: Anchor Press/Doubleday, 1974.

Shapley, Harlow. *Climatic Change.* Cambridge: Harvard University Press, 1965.

Toynbee, Arnold. *A Study of History.* New York: Oxford University Press, 1941.

Winkless, Nels III, and Browning, Iben. *Climate and the Affairs of Men.* New York: Harper's Magazine Press, 1975.

Wright, H. E., Jr. *Late Pleistocene Climate of Europe.* Boulder, Colorado: Geological Society of America, 1961.

Young, Patrick. *Drifting Continents, Shifting Seas: An Introduction to Plant Tectonics.* London and New York: Franklin Watts, 1976.

INDEX